AT HOME WITH THE HIGH ONES

A portfolio of photographs and text
by John S. Crawford

Published by
Alaska Northwest Publishing Company
Anchorage, Alaska

Printed in U.S.A.

First printing:
February 1974
Second printing:
October 1974

Library of Congress Catalog Card Number: 73-93156
International Standard Book Number: 0-88240-009-6

The golden eagle was soaring in widening spirals. In majestic solitude, his great wings set, he rode the chill wind of April above the northern Rockies. From his altitude he could see the entire Cascade Valley, the headwaters of the Panther, and eastward to the valley of the Ghost. It was too early in the spring for his succulent staples, the Columbian ground squirrel and the hoary marmot, to be out of hibernation. Perhaps, with the marvelous resolution of his vision, he was searching for a snowshoe hare—just beginning to change from winter white to the brown of summer—crouched nibbling in the willows; or, a mile or so up a confluent valley, for the telltale ravens and magpies near the carcass of a winter-killed caribou or elk.

But then as I watched from the rugged slope below, the eagle, seemingly without a wing beat, dropped in a beautiful, gliding sweep toward a band of bighorn sheep on the ridge.

I was puzzled, for only the occasional very young of mountain sheep fall prey to eagles. And these bighorns made up a mixed herd of adult rams and ewes and younger sheep down to sturdy yearlings. But I was about to find out that the golden is a fun-loving sportsman as well as something of an egotist.

Now the golden braked with flaring wings a few feet over the sheep, hovered momentarily, then climbed and banked around for another pass. Again he dropped, and, as he hovered, the bighorns looked up unperturbed, barely interrupting their grazing. But four more times the eagle planed down in wind-hissing dives, each time a bit lower. And then on the seventh pass, just as the golden flared his wings to hover, a young ram stood up on his hind hoofs like a rearing stallion. The ram's front hoofs blurred in jabbing thrusts. Brown plumage drifted away on the wind. The eagle staggered, side-slipped, nearly brushed the ridge crest, then flapped powerfully, recovered and climbed, apparently uninjured. Buoyed in a rising current, he set his wings and rode the wind out over the valley of the Ghost.

Since that memorable April morning a dozen years ago, above the Ghost River in the Palliser Range of the Canadian Rockies, I've seen golden eagles in sporting play with other bighorns, the Rocky Mountain goat and even the high-ranging grizzly. These encounters tell something of the proud and spirited nature of the high-dwellers—of both the eagles and the earthbound, and, to me, they are all among the most fascinating of the world's wild animals.

This book relates some of my experiences in studying the Rocky Mountain goat, the bighorn and Dall sheep and other wildlife linked in their mountain ecology. Every time that I've gone into mountain sheep or goat country I've learned something new about them, and there have been individuals among these mountaineers that I've been proud to consider my friends.

As far as trophy and meat hunting is concerned, I hung up my rifle to carry only cameras over a decade ago. Still, I have no argument with those who find an honest challenge in pitting their stamina against mountain storms and dangerous terrain to take a ram or billy. I don't begrudge the taking of any legal species as long as it is done within the rules of fair chase as defined by the Boone & Crockett Club, and as long as meat and hide are properly cared for and fully utilized. While I've met some slobs and certified S.O.B.'s among hunters afield, other big game hunters are among my best and most respected pals.

Many seasons afield have taught me that few wild animals die as painless a death as from a well-placed hunter's bullet. I recall an incident several autumns ago on Alberta's Big Smoky River. A young cow elk, attempting to ford at a bad place, had become mired down to her belly and brisket in quicksand-like muck a few yards out from the riverbank. She couldn't raise a hoof, but didn't drown, as she could keep her head, neck and upper shoulder area above the water. During the night a sharp freeze hit, forming about two inches of ice around the helpless elk. While the cow still had the energy to heave futilely against the entrapping mud, coyotes walked out on the ice and ate the head off her shoulders. There are many sides to Nature besides the bursting of buds in the spring, and that is one of them. And to condemn the hungry coyotes for helping themselves makes as much sense as condemning yourself for driving down to the market to purchase bacon and eggs.

Among outdoorsmen there has been much argument concerning the rarity of Rocky Mountain goats and mountain sheep mingling where they share the same range. I know quite a few veteran Canadian and Alaskan woodsmen who actually doubt that it ever happens. A couple of old-timers in Alaska once told me that goats and sheep aren't found on the same mountainside because they "don't like the smell of each other." This I can't believe. Once either species gets used to my presence on their range they don't even seem to mind how

I smell after three rugged weeks in the bush. Many times near mineral licks in the Canadian Rockies I've seen mountain goats and bighorn sheep within a few yards of each other. A friend who has made several trips into northern British Columbia's Cassiar district tells me of seeing the Stone sheep, *Ovis dalli stonei,* grazing as close as 50 yards from mountain goats.

I've never yet seen Dall sheep and mountain goats on the same mountainside, and haven't met anyone else who has; yet there are authentic reports of goat observations made within the boundaries of Mount McKinley National Park and in other Interior Alaska mountain areas that had been considered strictly sheep range. (In biological relationship, the wild mountain sheep of the world, though different in appearance—largely because of the short tail, and the stiff, hollow hair like that of the deer family, rather than wool fleece—and vastly different in temperament from their domestic relatives, are, nevertheless, true sheep. The Rocky Mountain goat, however, is misnamed; he is not a true goat as such, but an animal with characteristics of both the true goats and the Old World antelopes. He is a close relative to the chamois of the Alps and to the tahr, serow and goral of the Himalayas and other ranges of Asia.)

I doubt that either mountain sheep or mountain goats object to the presence of any other large ungulate species on their range. Though the two share much in common, ideal habitat requirements for mountain sheep and goats differ, and because of this the two species aren't too often found on the same slope, even in watersheds where there are large numbers of both, such as in the Brazeau and Big Smoky drainages of Alberta and in the Cassiar district of northern British Columbia.

In any mountain region of northwestern North America there is competition for the available grass and browse among several different big game species. In the Cascade Range of British Columbia and Washington, I've watched mountain goats feeding on alpine meadows within 40 yards of both blacktail deer and elk. Above the south fork of the Ashnola River, 25 miles or so from Keremeos, British Columbia, I have glassed large herds of mule deer and Rimrock bighorn sheep, *Ovis canadensis californiana,* grazing together on the open south slopes—already bare in early April while the snowpack lay heavy down in the timbered canyons. In some areas of the Canadian Rockies the encroachment of elk on the ancestral spring range of the Rocky Mountain bighorn is becoming a serious wildlife management problem. A classic example of this can be found on the slopes overlooking the Cascade River north of Banff, Alberta.

In the often close association between different species, it's interesting to observe how big game animals can get along, even, curiously, in situations of potential conflict.

One late November day in the Rockies of western Alberta, I was sitting with my back against a boulder, eating lunch on an Ice Age glacial moraine overlooking the valley of the Athabasca. The weather had moderated that morning to a comparatively pleasant eight above, after three days of sub-zero cold. Around 45 head of bighorns were also on the moraine, some within 15 yards of me.

I'd watched and studied this particular band for about a month. Approaching mountain sheep for close-up pictures is a completely different problem than stalking a ram with a rifle. To get good shots, even using telephoto lenses, you have to be much closer than the usual rifle range—and here there is just no substitute for time and patience. The eyesight of mountain sheep, vastly excelling that of man, is among the most highly developed in the animal kingdom, and there is little chance of your getting close enough for good pictures without their knowledge and acceptance of your presence.

Over a period of weeks this band of bighorns had become quite used to me, though any sudden or clumsy movement on my part would still spook them. I had a special interest in these sheep, because I was hoping to photograph a breeding season ram duel, and in this bunch there were five large mature rams of nearly equal body weight and horn growth.

Now on the moraine, most of the sheep were bedded down, while a few others were feeding on the tufts of grass that grew between the round, glacier-scoured rocks. One big full-curl ram was just standing still, looking off across the valley as classically posed and motionless as though he were a museum mount in a habitat group. Presently there was the clatter of rolling rock from over the edge of the moraine.

The sheep that I could see all turned their heads toward the sound, watching with interest, but with no indication of alarm. Those that were bedded down didn't stop chewing their cuds. Then up over the edge of the moraine walked a mule deer buck. Just moments before he'd crossed over the ice of the Athabasca and had swum the still open main channel. Icicles hung on his hide from neck to tail. On the muley's flank there was a red slash where another buck's antler tine had gouged a long, ugly furrow. Both the mule deer and bighorn rutting seasons were approaching their peaks, but this old boy appeared already too battle-weary to be looking for trouble.

The big four-pointer was apparently headed across the moraine to bed down in a stand of lodgepole pines. But he was walking slowly, head low, almost directly toward the ram.

I sat dead still, waiting, my back against the boulder, and if the buck noticed me at any time he gave no evidence of it. Of the sheep that I could see without turning, some kept watching the deer, others ignored him, and he paid no attention to them, just continued his slow walk across the moraine as though he knew the most direct route to his destination and was taking it, oblivious of the sheep. The big ram hadn't moved, and now it seemed certain that one or the other was going to have to give ground.

But neither did! The tines of the buck's left antler passed the ram's head hardly two feet away. The bighorn didn't move out of his tracks, just turned his head and, apparently with only mild interest, watched this bedraggled looking stranger walk by.

Country rugged enough for protection against predators is a basic habitat requirement for both mountain sheep and goats. Given this terrain, the choice of range in a particular watershed is determined by the feeding habits of each species. The mountain goat is both a grazer and a browser during the summer and early fall; in the winter he's primarily a browser. Mountain sheep—and on this point the Dall and the Stone to a larger extent than the bighorn—are essentially grazing animals the year-around. But there are some interesting geographical variations—from what might be considered most normal or typical—in the seasonal feeding habits of both mountain sheep and goats.

During the winter in areas of heavy, wet snow—such as in the coastal mountains of Southeastern Alaska and British Columbia, the British Columbia and Washington Cascades, and in a number of interior British Columbia ranges including the Kootenay Rockies, the Monashee Range and the Selkirks—goats descend into steep, cliffy areas, usually well below timber line, and subsist almost entirely on browse. In areas of the goat's range where there is dry powdery snow, as on the east slope of the Canadian Rockies—a specific example would be the DeSmet Range of the Alberta Rockies—goats winter on high ridges and slopes and feed on grasses and other alpine and subalpine plants exposed by the wind. They are found at elevations as high and sometimes higher than their usual summer range. The mountain goat's winter pelage is impressive anywhere you find him, but the coats of the animals that winter in the heights of the northern Rockies are magnificent. They are an environmental adaptation that protects the goats from lashing winds and temperatures that may drop lower than 60 below zero.

The Dall sheep, like the mountain goat in the northern Rockies, winters on wind-swept slopes and ridges. The Dall, unlike the bighorn, is rarely seen in valleys except when crossing over from summer range to winter range, and back again. When crossing a river valley, the Dalls tend to be very nervous, sensing their danger from predators; once a crossing is started—perhaps after hours of careful watching the route they will take, searching for any lurking hazard such as wolves or a grizzly—it is made swiftly, and the sheep do little, if any, feeding en route.

While some bighorns also winter in high country and forage on wind-exposed vegetation, most of them descend to lower slopes overlooking river valleys, but, even at low elevation, are rarely seen more than a few hundred yards from rugged terrain. In Alberta, during late November and December, I've frequently seen bighorns—young sheep and adults of both sexes—standing on the ice of frozen rivers browsing on muskeg willow along the banks.

Besides their diet staple of grasses, mountain sheep (here I'm including both the bighorn and Dall and their various subspecies) feed on sedges, mosses, heather, willow and the buds of various trees and shrubs. All of these items are found, too, in the mountain goat's fare, and the goat can do very well on coarser vegetation than mountain sheep prefer. Quite often I've observed goats browsing on the bough-tip growth of timber-line firs.

Mountain sheep and goats will travel miles, if need be, for salt, and the availability of mineral licks is a basic habitat requirement for both. Goats in the rugged coastal mountains of British Columbia, and in Southeastern Alaska goat country like the Boca de Quadra and Tracy Arm-Fords Terror areas, descend to salt water and lick ocean spray off the rocks.

Above the valley of Blue Creek, an upper confluent of Alberta's Snake Indian River (itself a tributary of the great Athabasca) rises an incredibly rugged and serrated mountain range known on the maps as The Ancient Wall. Spanning a watercourse on a lower slope of the range is a huge and majestic arch of limestone, carved by thousands of years of erosion. Around the base of this arch is a mineral lick where I've observed bighorns and mountain goats licking the clay-like soil within a few yards of each other. Imprinted in the moist earth there, I've seen the tracks of grizzly, cougar and wolf. For me, it is one of the most fascinating areas in the northern Rockies.

The most important predator in the ecology of the far northern sheep is the wolf. The Canadian bighorn faces the same predators the Dall and Stone do, plus the cougar. During some winters, particularly in the later portion from early February into April, unusually deep snow and generally severe weather conditions tend to favor the hunters over the hunted. During such times there is some predation on mountain sheep by the coyote, wolverine and lynx, as well as by the larger predators.

I don't feel it is appropriate, when considering the ecology of the north country, to call one animal an "enemy" of another. Predators are the classic opportunists; and, over the centuries, predation has been generally selective to the good of both predator and prey.

Healthy mountain sheep, except under unusual circumstances, would leave any predator far behind when fleeing toward rugged escape terrain. The sheep with a leg sprained while crossing a scree slide or with a heavy infestation of lungworms would be the one taken. Even when cut off from escape terrain—such as during a valley crossing from summer range to winter range—a healthy adult ram or ewe, if forced to make a stand, can be dangerous to an attacking predator.

Both predator and prey species are magnificent animals, thrilling to see in their habitat, and both are invaluable aesthetic assets to the north country scene.

I've seen coyotes, wolves and—on one memorable occasion—a cougar stalking bighorns, but I haven't yet seen any of these predators actually take a mountain sheep or goat, though I have observed wolves making kills on both moose and elk.

Early last March I was watching a band of bighorns on a steep slope overlooking the lower Snake Indian River in Alberta's Jasper National Park. Suddenly I noticed that several old ewes were standing taut and still, all looking toward an area of scattered pines upstream from them and slightly lower in elevation. Swiftly, through some way of communication, the attention of the entire band was drawn to that area. *This could be it,* I thought, as I focused a tripod-mounted 400mm lens on the slope.

I figured I just might be about to have another chance at photographing a cougar stalking bighorns. Raising binoculars and glassing the slope, I could see nothing moving among the trees. But as I lowered the glasses, I saw the sheep—reacting almost as a single individual—turn and race across a scree slide, bound up over ledges and stop on a rugged point overlooking the river. Then I spotted the predator that had spooked them; not a cougar, but an exceptionally large and heavy lynx that was suddenly skylined on the ridge. The cat stood still for a long moment, watching the sheep with evident great interest and, perhaps, disappointment. Then he turned back down toward the river, and, with stubby tail twitching and huge paws padding along the rocks, he faded almost magically back into the sparse cover of the upper slopes.

A curious experience happened a few years ago to a friend of mine on another upper Athabasca tributary. He was hiking one early November afternoon, accompanied by his big German shepherd dog, when he rounded a bend in the old trail and saw a band of bighorns near the riverbank just ahead of them. There were about two-dozen head, mostly ewes and lambs, with several young rams, and one powerful, heavy-horned, full-curl ram. The sheep had been grazing by digging and raking the snow away from the grass with their front hooves, and nibbling at willow buds. Now at the sight of the dog, they stood alert and still.

Suddenly the dog broke from the man—in his excitement ignoring the shouted orders of his master to come back—and raced toward the sheep. Immediately, at the rush of this wolf-like animal, the bighorns retreated, bounding up to the ledges of a high, steep scarp. All except the big ram.

For a few seconds the ram waited. Then he catapulted in a mighty lunge, and with an incredible burst of speed he charged the onrushing dog. The startled shepherd braked with stiff front legs, whirled and fled back toward his master. The chase lasted about 120 yards. The ram was closing fast, and a scant few yards from my friend he finally broke off the chase. The great bighorn turned slowly and reluctantly, walked a few yards up the slope, turned to eye them again belligerently, then continued climbing up to join his band. It was a wiser and thoroughly spooked German shepherd. And the dog wasn't the only one frightened. "Man!" my friend said, shaking his head as he told me about it, "I thought that guy was going to knock *me* into the river."

After years of observing both, I don't believe that any of the North American mountain sheep—not even the desert bighorn—can climb in as severe terrain as the Rocky Mountain goat. But in less difficult country the sheep are much faster than the goat—and faster, I'm sure, than deer. When really spooked they can cover the slopes with a graceful speed that is hard to believe even when you see it. The goat is a deliberate climber. I've watched goats carefully test rock points and ledges before putting their entire weight on them, in much the same way that an expert human rock climber does.

During the breeding season of the mountain goat, in late fall, the billy doesn't gather large harems. Four females with a billy is the largest number I've observed, and even this may be exceptional, with two or three the more usual number. The goat breeding season runs from early November into

December, and the young, usually singles but quite often twins, are born about six months later from about the last week in April through the first week or 10 days of June.

While fighting between mountain goat males (that is, serious rutting season combat) is much less spectacular than the duels between mountain sheep rams, it is far more dangerous to the antagonists. I've seen old nannies, and on a couple of occasions billies, resolve sudden arguments by charging each other and merely butting heads together. But billies in a serious fight circle each other, feinting and hooking with their black dagger horns, and these exchanges are sometimes lethal. The deadly effectiveness of the mountain goat's horns in defensive combat against predators is discussed later in this book. I watch the ritualistic spectacle of a classic bighorn ram duel with pretty much the same feeling that I have witnessing a great sporting event—say, the heavyweight boxing champion of the world against a top-rated challenger. But there is nothing to suggest sport in a grimly serious fight between two mountain billies, and it gives you a feeling of apprehension to watch it.

During a dozen autumns of observing bighorn breeding season behavior, the earliest I've seen the first restless aggressiveness of the rams has been late October, and in some years I've seen rutting activity extend into early January.

The bighorn's breeding season behavior is more complex than that of most of the other ungulates, especially regarding the fighting among rams. Early ram duels—during the last couple of weeks in October through the first week or so of November—will quite often establish a hierarchy of dominance among the rams of a particular band. In such cases, the dominant ram will sire most of the next spring's lamb crop. Unlike the bull moose and—to an even greater extent—the bull elk, which strive to maintain control over harems of cows by intimidating and fighting off challenger bulls, the mountain ram is interested in a ewe only when she is actually in season and receptive to breeding.

During several bighorn rutting seasons, I've made continuous observations of interesting situations where in one band of sheep there were several mature rams of almost equal body weight and horn growth. I've seen rams in such cases battle intermittently throughout the entire breeding season without any one of them emerging as a clear-cut victor. And, curiously, while the issue remains unresolved by fighting, often I've seen young rams—down to three and one-half years, the minimum breeding age of mountain sheep males—among the ewes, breeding, while several mature rams were battering heads less than 50 yards away! By the time he reaches his fourth autumn, the young ram is as interested in the rutting season activity as the mature rams, but at this age, of course, he is a competitor for the ewes only through stealth, not combat.

Several mature rams will often duel together, hitting each other in alternate charges. Younger rams frequently watch on the fringe of the action, but give ground quickly if challenged. Usually, in an active combat situation the young rams are largely ignored by the veterans.

Before real fighting gets under way, I've often observed rams bunched together, as if in conference. Then another ram, standing off several yards, may suddenly charge one of the group, driving a thudding blow into his ribs or flank, or, just as frequently, charging from behind, slamming and then levering up with his horns, and lifting the rump of his surprised and outraged colleague at a near 40-degree angle.

Such unsportsmanlike conduct may or may not precede a classic duel. But grunting and hoof-jabbing, with the two rams standing head to head, is the usual actual beginning of the drama. Following this preliminary ritual, the two rams abruptly turn from each other and walk away in opposite directions. Then, as if responding to an audible signal, both rams turn, stand up almost erect on their hind hoofs, facing each other on what could well be the most dramatic "line of scratch" in Nature, and start their charge. At the onset, both rams take several quick steps toward each other on their hind hoofs before each battler drops down in the final driving lunge toward his opponent. Just before collision—at a combined speed of 40-45 mph—the rams turn their heads on the side, thus taking the impact of a clean hit squarely on the boss of the horns. But if it isn't a clean hit, one or both rams can suffer an upper-jaw fracture. This is a fairly frequent occurrence, and more than half of the mature rams that I've seen showed definite evidence—in their humped nose-jaw profiles—of upper-jaw fractures sustained in combat.

Obviously, these fractures make it difficult—and sometimes for a few days impossible—for the injured rams to feed. But, with the rams' tremendous vitality, the healing process is rapid.

In cliff and canyon country where there is considerable echo effect, the sound of the rams' impact is somewhat like a rifle shot, but it still has a distinctively different tonal quality. For readers that engage in the good winter sport of curling, the closest approximation I've ever heard of the crashing impact of two mountain rams is to hold a pair of curling stones three or four feet apart and slam them together. Amplify the sound several times, and that's about it!

In the instant following the rams' impact, a shock wave ripples down the muscular back of each antagonist, and their hind hoofs leave the ground, driving out straight and stiff behind them.

Rams *do not* crash heads in swift exchanges. The overly cute editing of some motion pictures showing ram fights (timing the action to fit a musical background) has given this false impression. The force of the head-on collision is terrific, and following impact the combatants stand in front of each other for often 10 or 12 seconds, heads raised, stunned and absolutely motionless.

A duel between two mountain rams is one of Nature's classic dramas, but though horns are chipped and broken, noses frequently bloodied and upper jaws sometimes broken—it still lacks the sustained savagery of a well-matched bull elk fight.

Oddly enough, following bighorn duels, I've often seen the antagonists graze or bed down within a few feet of each other. Then, after a temporary truce to graze or rest, when a ram wishes to resume further combat, he approaches his feeding or bedded adversary and jabs or rakes his groin, belly or back with a front hoof, all the while cussing him out with the highly expressive growling grunts.

As with the mountain goat, the gestation period of both species of North American mountain sheep (there are several geographical races of both the bighorn and Dall species) is approximately six months. The lambs, singles or twins, are born in May and June. When the ewe nears her time to give birth, she climbs to a high, rugged, secluded area where she and the lamb have maximum protection from predators.

Following the rutting season, the mature rams, to a large extent, mingle with the ewes and young sheep on winter and spring range. During the summer and early fall, however, the rams keep to themselves, banding in groups of four or five up to herds numbering 30 head or more. I've seen the occasional old ram apparently ranging as a "loner."

Unless killed by predators, a fatal fall, an avalanche, pneumonia (mostly occurring in sheep already badly weakened with a lungworm infestation) or a hunter's bullet, a mountain sheep will live as long as his teeth last, rarely longer than 12 or 14 years. In the balanced ecology of a true wilderness situation, when the sheep's teeth become so worn he can't cut and chew the grasses that are his primary feed, it is highly unlikely that his death will result from actual starvation. In his gradually weakening condition, he will tip the scales of natural selection and fall easy prey to one of his hunters.

Though Dall sheep numbers have declined in some of the hunting areas more accessible to population centers in Alaska, they still occupy most of their original range. The less fortunate bighorn has been eliminated entirely from vast areas of his ancestral range in the United States. Overhunting in the early days, diseases carried by domestic sheep and competition on winter and spring range from domestic stock and from other wildlife—most seriously from elk—are all contributing factors.

There have been a number of attempts to restock the bighorn on areas of his former range from which he has been eliminated entirely or is in danger of extirpation. These restocking efforts have been made with the mutual cooperation of U.S. and Canadian federal, provincial and state wildlife agencies. The restocking attempts have been made with the Rocky Mountain bighorn, the desert bighorn and the California bighorn subspecies (mostly trapped in the Williams Lake area of British Columbia). Some of these transplants have been definite successes, others have apparently failed and with still others it's just too early to tell. I had the satisfaction a couple of springs ago of assisting in the capture and transfer of two-dozen head of Rocky Mountain bighorns from the Alberta Rockies to the Lostine River watershed in the Wallowa Mountains of northeastern Oregon.

On the Rocky Mountain goat, civilization has had relatively little impact. Today, through restricted hunting and comparatively inaccessible habitat, the mountain goat remains in geographical range and population probably very nearly the same as when he was first seen by white men. On no other large animal has human settlement had less effect.

The mountain goat is considerably less vulnerable to predators than the mountain sheep. This is partly due to his more precipitous habitat, but also because no animal lives that is more courageous in defensive combat. The goat is heavier than is generally believed, with some of the larger billies going well over 300 pounds. When attacked by predators, the goat's size, sinewy stamina, agility and lethal horns make him a formidable antagonist. The horns of adult goats of either sex average between 8 and 10 inches in length, with the record being a fraction over 12. As weapons, there is ample evidence of their effective use.

Old-time Canadian bushmen—big game guides, trappers and retired wardens—have told me of two cases where evidence in the form of remains showed that grizzlies had been killed by goats. Both were in the Alberta Rockies (I have heard of other similar cases in Alaska and British Columbia, but the evidence was not as well established). In one Alberta case, the goat, too, was killed. He had suffered a broken back from a paw blow after he had inflicted mortal wounds with his horns. The goat carcass with the broken back and the dead grizzly with the stab wounds through his vitals were found a few yards apart. In the other case, a slain bear was found with two obvious horn wounds, but the goat had either escaped injury or had at least been able to leave the immediate area.

A dozen miles north of the confluence of Alberta's Snake Indian and Athabasca rivers, a steep, black scarp rises some 300 feet from the west bank of the Snake Indian. This area is known as Shale Banks, and its bare, blackish soil is rich in various mineral compounds. This is one of the largest mineral licks in the northern Rockies and is used primarily by mountain goats and bighorns, and to some extent by mule deer, elk, moose and mountain caribou. In mid-June of '67 I witnessed there a dramatic example of a mountain goat's courage in a showdown with a predator.

On that particular afternoon, seven mountain goats—a young billy and three nannies with new kids—were in sight, scattered on the steep slope and licking the soil. As I looked at the upper reaches of the scarp, where it dropped away from a heavily timbered bench, a huge old billy walked out of the pines and began to work his way slowly along the top of the pitch. Then he stopped, and for a few moments licked at the rocky soil. Suddenly, he raised his head, looked behind him, and whirled in his tracks. His head went down, and his shoulder hump jutted like that of a bison. He stamped hard with a front hoof and pivoted slightly to the right so that he was facing away from me and toward the timber. Then in the pines a few yards from him, I saw a flash of tawny hide.

Standing on the east bank of the Snake Indian, I yanked my binoculars from their case, quickly focused and caught in full profile the sinewy, lethal grace of a cougar. Tail flicking, the great cat moved back and forth, seeming no more than six feet from the billy. The billy pivoted as the cat moved, always facing the cougar, head low, front hoof stamping in anger, black horns ready.

Again and again the cougar moved back just out of sight in the pines, only to reappear in a few moments as the drama went on.

For 25 minutes a formidable and determined predator and his powerful and courageous prey held each other to a stand-off.

Finally the cougar faded back into the pines, and minutes passed without his return. The great shaggy billy turned, traversed a few yards to his chosen route and picked his way unhurriedly down the long pitch to the river. At the time the Snake Indian was high, near the peak of snow-melt and glacier runoff. Placing his hooves in precarious niches in the rock, the billy was almost standing on his head, waves slapping him in the face, as he drank deeply from the wild, frigid torrent of the Snake Indian.

Mountain goats, in contrast to the rather high-strung mountain sheep, are calm by nature and often display remarkable composure when encountered in their rugged uplands. But when they descend out of their usual habitat, goats—and particularly females with young—tend to be spooky when they drop down to visit mineral licks or to cross valleys to another range.

In early July of '67, I was camped on a glacier moraine flat, just north of the Scott Glacier in the upper Whirlpool Valley in the wilderness backcountry of Jasper National Park. This remote area is a few miles east of the Continental Divide and the Alberta-British Columbia boundary. At the time, I was studying the mountain grizzly and investigating reported wolf denning sites in the upper Whirlpool Valley. I hiked many miles in the watershed from the Whirlpool—Athabasca confluence westward up the Whirlpool Valley over Athabasca Pass and across the divide into British Columbia. One morning I woke just before sunrise to a clatter of rolling rock, loud splashing in the river, then more clattering of rock just outside my tent.

My first drowsy thought was, *maybe one of Don's horses threw his hobbles and lit out* (at the time Warden Don Rose was building a log footbridge over a lower fork of the Whirlpool). I rolled out of the sleeping bag and pulled up the entrance flap of the mountain tent. A thoroughly wet nanny and kid had just forded the river and were now running across the round, glacier-smoothed rocks just in front of my tent. When I first saw them they were less than 20 feet from me. The goats had been running hard, and they paused momentarily to look at me, but took off again immediately, the nanny running in a bobbing, shoulder-heaving, bison-like gait, the kid right at her side. I hurried out of the tent to watch.

There is a huge moraine flat across from the Scott Glacier, and the goats had fully a mile to go before they were in what, for them, would be safe terrain. With almost incredible stamina the goats never let up their pace, running until they were out of sight in the cover of the steep, timbered slope on the north side of the valley. *If there's even a coyote ranging that slope right now,* I thought, *she's in trouble. She might get just too bushed to fight.*

The Rocky Mountain goat is essentially a herd animal. They band together in from small groups of three or four goats up to herds of several dozen. In the northern Cascade Range near the British Columbia-Washington boundary I once counted 64 goats in a band. I've heard of still larger bands in reports from Canadian and U.S. forestry agencies and national park services. The large herds of goats I've observed were composed of adults and younger goats of both sexes, and kids born that spring.

In every part of the mountain goat's range that I've worked in—including goat country in Montana, Idaho, British Columbia, Washington, Alberta and Alaska, I've observed lone goats. With few exceptions these were large old males.

Probably some of these old boys just simply prefer their own company, others may have been driven out of their bands by younger males and some may have been temporarily separated from their herds while recovering from injuries.

Late one August afternoon, on a flank of Bismarck Peak in the Washington Cascade Range, a lone billy and I saw each other at the same time less than 30 yards apart. Instead of showing the usual curiosity and composure, he whirled and raced across a scree slide, not stopping once to look behind him, until he disappeared behind a granite shoulder 300 yards away. I still wonder if he might have taken a hunter's bullet a season or two before, recovered and associated the sight and scent of man with the lingering pain of his injury. Judging from most of my experience with goats, his totally spooked reaction was highly unusual.

Curiosity is a basic ingredient of the mountain goat's personality. On a frosty September morning in the upper Skagit country near the Washington-British Columbia border, I suddenly woke—not drowsily, but, curiously, totally alert—at the first grey of daybreak in the east. I wasn't using a tent, and, lying warm in the good down sleeping bag, listening to the chill wind in the alpine firs and the distant fall of water in a chasm, I thought how odd it was that I had awakened so early, for the previous day had been long and tough. But a moment later I felt several soft nudges against my feet and lower legs. For a few seconds I lay completely still, then very slowly raised my head. There was a blur of white coats in the dusky light, and the clacking of hooves on rock. Then a mother goat and two kids stopped about 20 yards away and stood looking back at me. For some time they had been sniffing me as I lay asleep!

A few moments later the mother grunted what seemed to be an order, and the family began climbing away. As they turned, one of the kids, perhaps in answer to his mother, made a curious, plaintive sound that I'd heard many times before. Its tonal quality was something between the bleat of a lamb and the mew of a kitten, but sustained without much rise and fall in inflection. The kids use a long version of this call to indicate distress, though, like the complaints of youngsters of any species, it's not necessarily of great importance. If you're near a band of goats when the kids are being weaned, you'll hear it almost constantly. A kid will make this sound when momentarily stopped while traveling, when he thinks he can't make that next big step to a higher ledge. If a band of goats is in timber-line cover, this call of the kids may be your first indication of their presence, for if the wind is right the sound will carry several hundred yards.

Goats are almost incredibly sure-footed, but they do occasionally fall. The kids, born with the courage and coordination of their parents, must learn their own limitations. I've seen four kids fall, but not far, and none seemed more than badly bruised. They just stood up from the ledge or rock slide, made the curious, mewing bleat, looked things over again and then climbed on up, sometimes taking a different route. In the Selkirk Range of British Columbia I found one adult carcass that was clearly a fall casualty, not an avalanche victim.

That goats don't fall more often is still a marvel to me. In the British Columbia Monashee Range I watched a billy moving across a granite pitch so steep that it seemed a fantastic feat for him just to be on it. He stopped frequently to nibble at what appeared to be lichen growth, sometimes reaching for meager tidbits several feet above him by standing up on his hind hoofs. Once, with appalling nonchalance, 300 sheer feet above a jumble of shattered slabs, he raised a hind hoof and thoroughly scratched his back!

Anywhere in the mountain goat's range, he can be seen with more predictable success than any other big game animal—with the exception, perhaps, of bears on a major salmon spawning stream. For persons physically incapable of foot travel in rugged terrain, an easy and enjoyable way of seeing mountain goats is by boat trip into one of the mainland inlets of coastal British Columbia and Southeastern Alaska. In Southeastern Alaska, the Tracy Arm-Fords Terror area, Endicott Arm, Cleveland Peninsula and Boca de Quadra south of Behm Canal are recommended as particularly good goat observation areas.

For seeing goats at close range during the summer and early fall, you should generally figure on hiking up to snow line. Goats love to play and rest on glaciers and snowfields for relief from summer heat.

Late in July in the northern Cascade Range, I once climbed up to a band of 27 mountain goats lying on a shrinking snowbank in a saddle between two peaks. Just about all the available space on the snow was occupied. As I eased cautiously up toward them to try for still closer photos, the goats stood up and moved off leisurely in a single-file climbing traverse up a rugged slope bordering the saddle. Closely inspecting their bedding area, I found that the lanolin from the goats' hair gave the snow a slick, oily feeling to the touch. The snowbank was also marked by a very definite—but not unpleasant—odor, a scent no more "gamey" than that of elk.

Only the eagle can follow the mountain goat in his toughest terrain, and on rare occasions the eagle does manage to catch a young kid far enough away from his fiercely defensive mother. But snow avalanches in the spring and late fall, rather than any predator, are the greatest single hazard the goat faces. Often the first meat the mountain grizzly feeds on after coming out of his semi-hibernation is an avalanche-killed goat, and he will excavate tons of rock and snow to get at a carcass.

Back in '61 I made an early April trip into the British Columbia Selkirks to photograph the mountain goat, the grizzly and the spectacular and deadly spring avalanches. I was working out of the Glacier Station area about midway between Revelstoke and Golden, British Columbia. One memorable day, I was making a snowshoe traverse, climbing toward Rogers Pass, above the snowbound grade of the last unfinished link of the Trans-Canada Highway. It was a pleasant morning, about 60 degrees, sunny, but with a misty overcast shrouding the great peaks overlooking the Rogers summit.

Nearing the pass, I sighted eight mountain goats browsing on willow tips high on the rugged slope above me. Several stopped feeding a few moments to watch me. Without thinking about it, I raised a hand in greeting, just deeply feeling the warmth of meeting again an old and respected friend.

Minutes later I dropped down to the grade at the summit of the pass. The snow was more settled and fairly firm here, so I unstrapped and kicked out of the webs. I felt a slight wind against my face, and I looked up at the overcast that was beginning to move across the upper slopes. Then, as I watched the clouds, I heard a sound from far above me. It was a muted rumble like distant thunder, or a flight of jet fighters at high altitude.

Lost in the spooky, wreathing clouds of the heights, the sound was building. It was the start of an avalanche. The late morning sun had broken the delicate balance between friction and gravity; now thousands of tons of snow and timber debris were gathering speed on the face of the mountain. It now sounded almost directly above me, but I wasn't ready with the cameras and there wasn't going to be time to get ready.

Moments later through the shifting mist, I saw it, coming in a crashing swath like storm breakers on a beach. Then, well below my direct line of vision, I just caught another movement. It was the goats. Uncasing binoculars, I picked them up in the field. They were in a traversing climb, going as fast as I've ever seen goats move, but *toward* the apparent route of the slide. Then I saw and realized where they were headed, to a great quartzite buttress that jutted from the slope.

A cross wind was pushing the mist, and in a few seconds the goats were lost from view. The sound of the avalanche had grown to a roar. The slide hit a pocket, geysered up like an explosion, poured over the rim, and funneled into a steep ravine. Beside this gorge, partially obscured in mist, stood the great shoulder of rock.

The overcast thinned and swirled in the wind, briefly parted, and there were the goats. Eight of them stood on successive ledges, safe but scant feet from the slide. In a tableau of timeless drama and beauty they stood still, calmly watching the roaring, plunging maelstrom of snow.

Gripped by the impact of the scene, I stood still. But the thought finally came, *get going, she's coming all the way!*

I grabbed up the snowshoes. Maybe a guy can't sprint with webs under his arm, pack on his back and binoculars flopping on his chest, but I did the best I could.

The avalanche boomed across the road grade, plunged into the Stoney Creek gorge, billowed into clouds as it hit the creek bed and thundered up the opposite slope on its momentum.

Moments later, when the last rumbling echoes were stilled, I walked back and looked at the great ridge of snow and debris across the grade—snow compacted by the velocity of the avalanche, laced with boughs, bark and shards of splintered fir trunks. I looked up at the great stone buttress. The goats were moving off their ledges and calmly starting to browse again. For a few moments, two young billies butted each other in playful sham combat. Perhaps man, with his neuroses from this dangerous age, can learn something from mountain goats!

In late afternoon after photographing several smaller slides, I tightened my web bindings, and headed back down toward the valley of the Illecillewaet River and Glacier Station. I was still thinking about the goats. From the sound in the high overcast, they couldn't have known just where the slide was coming. But they did know one safe place on that face of the mountain. They had known where to go, and, with the composed courage of their kind, they had watched as the greatest danger of their lives passed within a few feet of them.

I stopped a few moments and looked up at them, browsing on the slope, big and shaggy in their full winter coats.

Take care of yourselves, I thought.

John S. Crawford, the author of *At Home With The High Ones,* grew up near Sunnyside, Washington. He formerly worked for the U.S. Fish & Wildlife Service, and is a graduate of the University of Washington where he studied communications and zoology. He served in the U.S. Marine Corps, and later as a survival instructor on the Field Medicine & Surgery Committee, U.S. Army, at Fort Sam Houston, Texas. Mr. Crawford is now a freelance outdoor author, nature cameraman and wildlife researcher, and is currently engaged in a study and photo project on wolves in the Canadian Rockies.

His photographs of northern wildlife are used by the U.S. Forest Service, the U.S. National Park Service, and in Canada by the National and Historic Parks Branch of the Department of Indian Affairs and Northern Development.

Among other magazines and newspapers, his articles and/or pictures have appeared in ALASKA® magazine, OUTDOOR LIFE, TRUE, FIELD & STREAM, SUNSET, NATIONAL GEOGRAPHIC, ALASKA GEOGRAPHIC, NATIONAL WILDLIFE, THE SEATTLE POST-INTELLIGENCER and THE SEATTLE TIMES. His wildlife photos have been published by the nature book departments of the READER'S DIGEST, TIME-LIFE, McGRAW-HILL, LITTLE-BROWN, LANE MAGAZINE & BOOK CO. and CHANTICLEER PRESS.

Mr. Crawford's photographic work on the wildlife of the northern Rockies is represented in *The Athabaskan Peoples: Strangers of the North,* an exhibition produced jointly by the National Museum of Canada and the Royal Scottish Museum, to be shown in North America and Europe and opening in summer 1974 in Edinburgh, Scotland.

LIST OF PLATES

Plate 23—The lunge just before impact.

Plate 24—Bighorns weighing about 275 pounds each take a combined impact of about 45 mph.

Plate 25—A bighorn ram with horns chipped from combat.

Plate 26—Rocky Mountain bighorn rams, overlooking Athabasca Valley, Alberta.

Plate 27—Rocky Mountain ewe and three-week-old lamb, Palliser Range, Alberta.

Plate 28—Yearling bighorn lambs, DeSmet Range, Alberta Rockies.

Plate 29—Bighorn ram bedded down on a bench above Athabasca Valley.

Plate 30—Bighorn herd composed of ewes, rams and young sheep, overlooking Moosehorn Creek, Alberta Rockies.

Plate 31—Spring avalanche, Kootenay Range, British Columbia Rockies.

Plate 32—The Ancient Wall bordering the valley of Blue Creek, Alberta Rockies.

JOHN S. CRAWFORD

PLATE 1— COYOTE

JOHN S. CRAWFORD

PLATE 2 — NORTHERN GRAY WOLF

JOHN S. CRAWFORD

PLATE 3 — ROCKY MOUNTAIN COUGAR

PLATE 4—MOUNTAIN GRIZZLY, ALASKA RANGE

JOHN S. CRAWFORD

PLATE 5 — GRIZZLY BEAR

JOHN S. CRAWFORD

JOHN S. CRAWFORD

PLATE 6 — CANADA LYNX

JOHN S. CRAWFORD

PLATE 7 — CANADA LYNX FEEDING ON A DOE DEER

PLATE 8 — HOARY MARMOTS

JOHN S. CRAWFORD

JOHN S. CRAWFORD

PLATE 9 — ROCKY MOUNTAIN BILLY GOAT

JOHN S. CRAWFORD

PLATE 10 — ROCKY MOUNTAIN GOATS

JOHN S. CRAWFORD

PLATE 11—TWO-MONTH-OLD ROCKY MOUNTAIN KID

JOHN S. CRAWFORD

PLATE 12—ROCKY MOUNTAIN GOATS, BRITISH COLUMBIA

JOHN S. CRAWFORD

PLATE 13 — ROCKY MOUNTAIN GOAT IN FULL WINTER PELAGE

JOHN S. CRAWFORD

PLATE 14—DALL SHEEP RAMS, NORTH SLOPE, ALASKA RANGE

JOHN S. CRAWFORD

PLATE 15—HERD OF DALL SHEEP EWES AND LAMBS, NORTH SLOPE ALASKA RANGE

JOHN S. CRAWFORD

PLATE 16 — DALL SHEEP RAMS

JOHN S. CRAWFORD

PLATE 17 — DALL SHEEP

JOHN S. CRAWFORD

PLATE 18—DALL RAMS

JOHN S. CRAWFORD

PLATE 19 — ROCKY MOUNTAIN BIGHORN RAMS, MIETTE RANGE, ALBERTA ROCKIES

JOHN S. CRAWFORD

PLATE 20 — BIGHORN RAMS IN CLASSIC POSTURE AT START OF COMBAT

JOHN S. CRAWFORD

PLATE 21 — BIGHORNS PREPARING TO DUEL

JOHN S. CRAWFORD

PLATE 22—STANDING PARTIALLY ERECT. BIGHORNS BEGIN TO CHARGE

JOHN S. CRAWFORD

PLATE 23 — THE LUNGE JUST BEFORE IMPACT

JOHN S. CRAWFORD

PLATE 24 — BIGHORNS WEIGHING ABOUT 275 POUNDS EACH TAKE A COMBINED IMPACT OF ABOUT 45 MPH.

JOHN S. CRAWFORD

PLATE 25— A BIGHORN RAM WITH HORNS CHIPPED FROM COMBAT

JOHN S. CRAWFORD

PLATE 26 — ROCKY MOUNTAIN BIGHORN RAMS

JOHN S. CRAWFORD

PLATE 27—ROCKY MOUNTAIN EWE AND THREE-WEEK-OLD LAMB

JOHN S. CRAWFORD

PLATE 28 — YEARLING BIGHORN LAMBS, DE SMET RANGE, ALBERTA ROCKIES

JOHN S. CRAWFORD

PLATE 29 — BIGHORN RAM BEDDED DOWN

JOHN S. CRAWFORD

PLATE 30— BIGHORN HERD COMPOSED OF EWES, RAMS AND YOUNG SHEEP

JOHN S. CRAWFORD

PLATE 31— AVALANCHE, KOOTENAY RANGE, BRITISH COLUMBIA ROCKIES

JOHN S. CRAWFORD

PLATE 32 — THE ANCIENT WALL BORDERING THE VALLEY OF BLUE CREEK, ALBERTA ROCKIES

BILDNER

Wo&Wie

Microsoft Office

Schnelle Tasten für Word und Excel

Diese Zusammenstellung besonders häufig benötigter und nützlicher Tastaturbefehle soll Sie bei Routineaufgaben in Word und Excel unterstützen. Bis auf wenige Ausnahmen können die Tastenkürzel in allen Office-Versionen sowie Microsoft 365 verwendet werden.

Beachten Sie, dass nicht auf jeder Tastatur alle gängigen Zeichen tippbar sind. So können je nach Tastatur-Layout bspw. die Tasten für Ä, Ö, Ü, ß oder auch der Ziffernblock fehlen.

Office-übergreifende Tasten

Dokumente verwalten

Neues Dokument	Strg + N
Dokument öffnen	Strg + O
Dokument speichern	Strg + S
Dokument schließen	Strg + W oder Strg + F4
Dokument drucken (Dialogfenster Drucken)	Strg + P
Seitenansicht, bzw. Druckvorschau (2003)	Strg + F2
Anwendung beenden / Fenster schließen	Alt + F4
Fenster maximieren, bzw. Wiederherstellen der vorherigen Größe	Strg + F10
Wechseln zwischen dem deutschen und US-Tastaturlayout	Alt + ⇧

Allgemein

Kopieren des markierten Textes oder Objekts	Strg + C
Ausschneiden des markierten Textes oder Objekts	Strg + X
Einfügen aus Zwischenablage (Text oder Objekt)	Strg + V
Markiertes Objekt duplizieren (wird anschließend sofort eingefügt, ohne Zwischenablage)	Strg + D
Rückgängigmachen der letzten Aktion	Strg + Z
Wiederholen der letzten Aktion	Strg + Y oder F4
Markierte Auswahl fett formatieren	Strg + ⇧ + F
Markierte Auswahl kursiv formatieren	Strg + ⇧ + K
Markierte Auswahl unterstreichen	Strg + ⇧ + U

ZugriffstastenInfos

Abgesehen von den typischen Shortcuts bieten die Office-Programme eine weitere Möglichkeit, um mit schnellen Tastenkombinationen viel Zeit zu sparen: Die sog. ZugriffstastenInfos. Diese lassen sich durch Drücken der Alt-Taste aktivieren. Daraufhin erscheinen über den Funktionen im Menüband kleine Felder mit Tasten. Durch Drücken der entsprechenden Zeichen können Sie die Funktion aufrufen bzw. weitere Felder für Unterfunktionen anzeigen. Sehr praktisch!

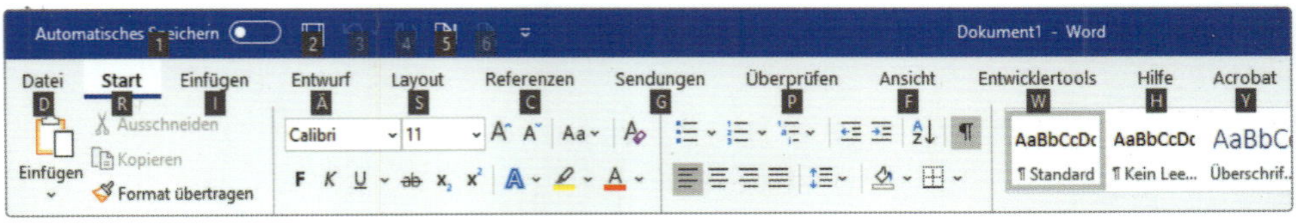

Microsoft Word

Allgemein

Einfügen Absatzende	↵
Einfügen Zeilenumbruch, kein Absatzende	⇧ + ↵
Eingabe geschütztes Leerzeichen	Strg + ⇧ + Leertaste
Eingabe geschützter Bindestrich	Strg + ⇧ + -
Eingabe bedingter Trennstrich	Strg + -
Einfügen Seitenumbruch	Strg + ↵
Einfügen Spaltenumbruch	Strg + ⇧ + ↵
Einfügen Datum (aktualisierbar)	Alt + ⇧ + D
Einfügen Seitenzahl	Alt + ⇧ + P
Anzeigen nicht druckbarer Zeichen	Strg + ⇧ + +
Zeichen links von der Einfügemarke löschen	⇦
Zeichen rechts von der Einfügemarke löschen	Entf
Wort links von der Einfügemarke löschen	Strg + ⇦
Wort rechts von der Einfügemarke löschen	Strg + Entf

Positionieren

Ein Zeichen nach links	←
Ein Zeichen nach rechts	→
Ein Wort nach links	Strg + ←
Ein Wort nach rechts	Strg + →
Ein Absatz nach oben	Strg + ↑
Ein Absatz nach unten	Strg + ↓
Eine Zeile nach oben	↑
Eine Zeile nach unten	↓
An das Zeilenende	Ende
An den Zeilenanfang	Pos 1
An den oberen Rand des Fensters	Alt + Strg + Bild ↑
An den unteren Rand des Fensters	Alt + Strg + Bild ↓
Eine Bildschirmseite aufwärts (Bildlauf)	Bild ↑
Eine Bildschirmseite abwärts (Bildlauf)	Bild ↓
An den Anfang der nächsten Seite	Strg + Bild ↓
An den Anfang der vorherigen Seite	Strg + Bild ↑
An das Ende des Dokuments	Strg + Ende
An den Anfang des Dokuments	Strg + Pos 1
Zur letzten Bearbeitungsstelle	⇧ + F5
Nach dem Öffnen eines Dokuments zur zuletzt bearbeiteten Stelle	⇧ + F5

Microsoft Excel

Allgemein

Eingabe in Zelle abschließen (Zelle darunter wird markiert)	↵
Eingabe in Zelle abschließen (Nächste Zelle rechts wird markiert)	⇆
Eingabe in Zelle abbrechen	Esc
Neue Zeile in Zelle beginnen	Alt + ↵
Eingabe in den gesamten markierten Zellbereich überneh-men	Strg + ⇧ + ↵
Nächste Zelle rechts / links markieren	→ / ←
Nächste Zelle oben / unten markieren	↑ / ↓
Markierung nach rechts / links erweitern	⇧ + → / ⇧ + ←
Markierung nach oben / unten erweitern	⇧ + ↑ / ⇧ + ↓
Erste Zelle in Zeile markieren	Pos 1
Erste Zelle im Arbeitsblatt (A1) markieren	Strg + Pos 1
Gesamtes Arbeitsblatt markieren	Strg + A
Wenn eine Zelle innerhalb einer Tabelle markiert ist, dann zweimal Strg + A drücken	
Aktuelle Spalte markieren	Strg + ␣
Aktuelle Zeile markieren	⇧ + ␣
Auswahl nächstes / vorheriges Arbeitsblatt	Strg + Bild ↑ / Strg + Bild ↓
Zellen einfügen (Dialogfenster) anzeigen	Strg + +
Zellen löschen (Dialogfenster) anzeigen	Strg + -
Neues Arbeitsblatt einfügen	⇧ + F11

Eingabe

Aktuelles Datum einfügen	Strg + .
Aktuelle Uhrzeit einfügen	Strg + ⇧ + : (Doppelp.)
Markierte Zelle bearbeiten (Bearbeiten-Modus) Einfügemar-ke erscheint in der Zelle	F2
Bearbeiten-Modus: Cursor an den Anfang	Pos 1
Bearbeiten-Modus: Cursor an das Ende	Ende
Kopiert den Inhalt aus der darüberliegenden Zelle in die aktuelle Zelle	Strg + ⇧ + ,
Unten ausfüllen (Format und Inhalt der ersten Zelle des mar-kierten Bereichs in die darunterliegenden Zellen kopieren)	Strg + U
Rechts ausfüllen (Format und Inhalt der ersten Zelle des mar-kierten Bereichs in die angrenzenden Zellen rechts kopieren)	Strg + R
Suchen Dialogfenster öffnen	Strg + F
Ersetzen Dialogfenster öffnen	Strg + H

Schriftformatierung

Öffnen des Dialogfensters Schriftart	Strg + D
Vergrößern des Schriftgrads um 1 Punkt	Strg + 9
Verkleinern des Schriftgrads um 1 Punkt	Strg + 8
Vergrößern des Schriftgrads	Strg + ⇧ + <
Verkleinern des Schriftgrads	Strg + <
Groß-/Kleinschreibung ändern	⇧ + F3
Als Großbuchstaben formatieren	Strg + ⇧ + G
Unterstreichen nur von Wörtern (keine Leerzeichen)	Strg + ⇧ + W
Doppelt Unterstreichen	Strg + ⇧ + D
Kapitälchen	Strg + ⇧ + Q
Tiefgestellt (automatischer Abstand)	Strg + #
Hochgestellt (automatischer Abstand)	Strg + +
Alle Zeichenformatierungen entfernen	Strg + ␣
Textformatierung kopieren (Format übertragen)	Strg + ⇧ + C
Der Markierung kopierte Formatierung zuweisen	Strg + ⇧ + V
Copyright-Symbol einfügen (©)	Alt + Strg + C
Symbol für eingetragene Marke einfügen (®)	Alt + Strg + R
Markensymbol einfügen (™)	Alt + Strg + T

Absatzformatierung

Einfacher Zeilenabstand	Strg + 1
Doppelter Zeilenabstand	Strg + 2
1,5-facher Zeilenabstand	Strg + 5
Hinzufügen / Entfernen eines einzeiligen Abstands vor einem Absatz	Strg + 0 (Null)
Linksbündig ausrichten	Strg + L
Rechtsbündig ausrichten	Strg + R
Im Blocksatz ausrichten	Strg + B
Zentriert ausrichten	Strg + E
Linken Einzug vergrößern (jeweils 1,25 cm)	Strg + M
Linken Einzug verkleinern (jeweils 1,25 cm)	Strg + ⇧ + M
Hängenden Einzug vergrößern	Strg + T
Hängenden Einzug verkleinern	Strg + ⇧ + T
Alle Absatzformatierungen entfernen	Strg + Q

Formatvorlagen zuweisen

Formatvorlage Standard zuweisen (entfernt alle Formatierungen)	Strg + ⇧ + N
Formatvorlage Überschrift 1 zuweisen	Alt + 1
Formatvorlage Überschrift 2 zuweisen	Alt + 2
Formatvorlage Überschrift 3 zuweisen	Alt + 3